Northern Bee Books
Scout Bottom Farm
Mytholmroyd
Hebden Bridge
HX7 5JS (UK)

www.northernbeebooks.co.uk
Tel: 01422 882751

© 2018

ISBN: 978-1-912271-37-5

The
Honeybee
in Focus

Written and illustrated

by

W.M.Sweeney

For Ian, Robert and Emma

Contents

Preface

My book is about beekeeping and honeybees in general. There
are six chapters, the first three dealing with the castes.

1. The Worker's Roles.

2. The Queen.

3. Drones.

The fourth chapter is concerned with honey bee behaviour.

4. Behaviour

No bee book can be written without a look at pests and disease and how to avoid/treat them.

5. Pests and Diseases.

Finally, I have included a chapter about the beekeeper and some things
which may make life easier and healthier for the bees.

6. Beekeeping

With chapters like this the book appears to be written as most other beekeeping
tomes. This one is completely different in that I have attempted to convey the facts and
nuggets of helpful information by means of what I hope are hilarious cartoons.

Beneath each cartoon will be a "Caption" - serious facts to explain each cartoon,
thus opening up the book to beekeepers and non-beekeepers alike.

I see this as a way to entertain beekeepers and to engage non-beekeepers, young and
old, who are interested in bees. Hopefully, I may encourage them into finding out more.

Chapter 1
The Worker's Roles

> **"** *All the hive's a stage. And all the workers merely players. They have their exits and their entrances. Each bee in her time plays many parts.* **"**

Shamelessly plagiarised from "As You Like It" by William Shakespeare

Chef

From an early age bees will produce brood food from their hypopharangeal glands to feed younger bees.

Returning from foraging they will routinely give "samples" of nectar to other workers as part of their direction to good food sources.

Cleaner

Immediately after emerging from its cell in the comb, a new bee will begin to clean.

Preparing the empty cell, ready for the queen to lay an egg.

Dancer

The "Waggle dance" is a fascinating and complicated series
of manoeuvres performed regularly in the hive.

It is a means of communicating to other bees the exact location of forage.

Food Delivery and Storage

Foraging, delivery and the safe storage of food is vital to a hive if it is to survive the winter.

This is when little further food is to be had. Water, nectar and pollen are all brought to the hive.

Pollen is taken from many different sources as bees need a
variety or nutrients, hence the variety of colours.

Heating Engineer

Worker bees control the temperature of the hive by rapidly fanning their wings. Ideally the brood nest should be between 32 and 35 degrees in order that brood can be raised.

Fanning is also used to regulate the hive humidity. Water is brought into the hive by foragers and can be used in this process.

Hive of Activity

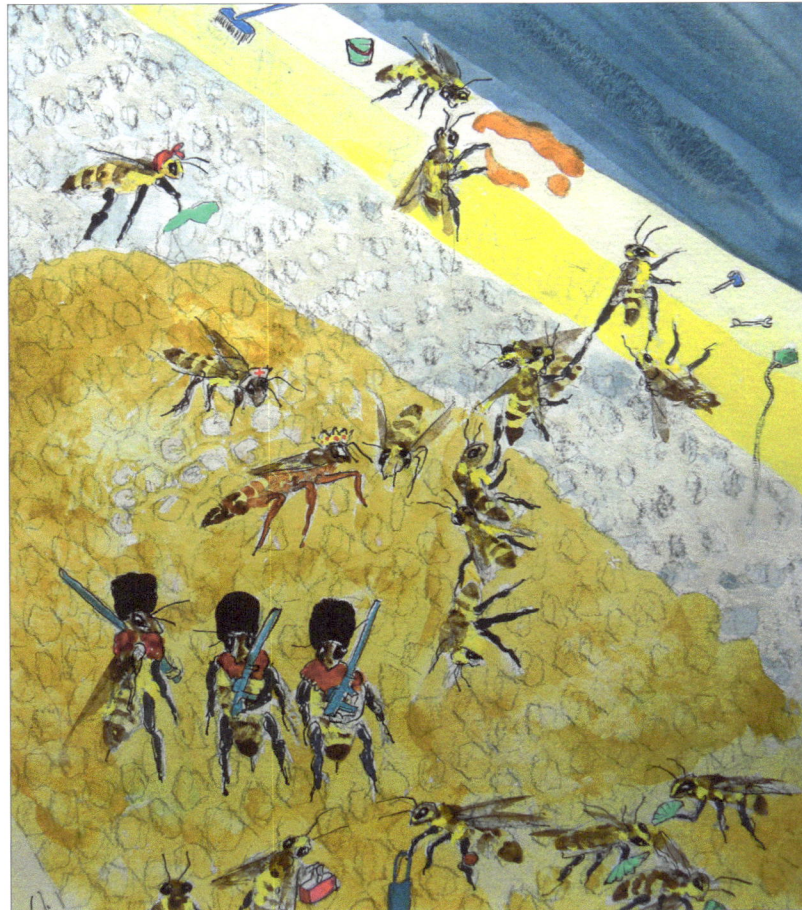

Within the hive, each bee has a designated task which it performs until it is needed to switch on to another.

There are breaks and rests, although hardly noticeable. The foraging bee will literally work itself to death beyond the hive.

Nurse

Nursing new bees is a task performed by very young bees. They feed and groom the young.

Brood food is a mixture of nectar and pollen, bound with secretions
from the hypopharangeal gland of the nurse.

Queen Attendant

Once mated, the queen does nothing but lay eggs. Workers will usually, at some point, take a turn in her retinue.

This will involve cleaning and grooming her, feeding her the best food. They will check on her constantly, never leaving her alone.

It is vital that she must be at her peak, if the colony is to thrive.

Forager

Knowing which forage bees are taking is invaluable to the beekeeper as
it determines the character of honey and other bee products.

Confusingly each plant has three main names. The universal latin (Biological name).
The common English name and the one I like to call "Ancient slang".

This is the rather more colourful one and the one I always seem to remember.

Soldier

Older bees, usually 12-18 days, will perform the task of defending the hive. During an attack, the alarm pheromone will alert other bees. Stinging is the main defence. Other bees will try to sting in the same spot at the sting pheromone is released. Smaller attackers will be "Balled"(Covered with bees) until overheated and smothered.

Against a deadly prey like an Asian hornet, smaller colonies have little chance. The beekeeper is the only hope of that colony.

Undertaker

The sting of a honeybee worker is barbed and will usually remain in the thick skin of a human where it continues to pump venom.

If the sting is removed immediately, there is minimal effect. The bee, however will die.

Any bee dying in the hive will be removed by undertaker bees to prevent disease.

Sculptor

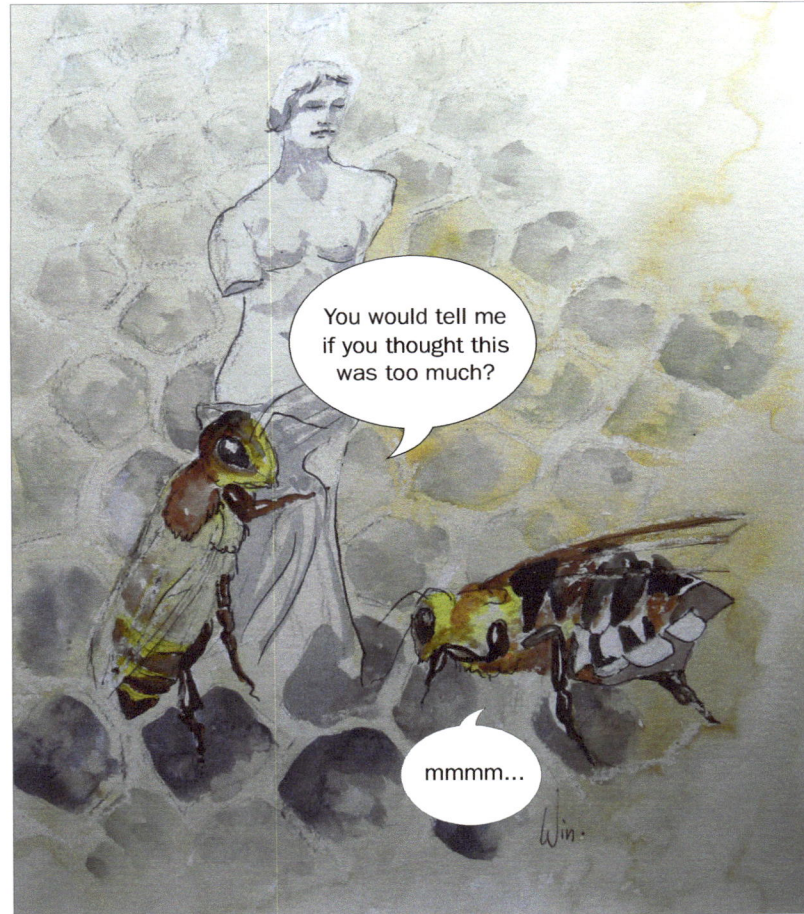

Perhaps honeybees did not fashion the Venus De Milo but the comb they build from the wax platelets extruded though segments of their bodies is no less spectacular.

Chapter 2

The Queen

" *No queen has clean hands.* **"**

George R.R.Martin, A Dance with Dragons

Birthday

A queen honeybee can live up to five years, most usually three to four, although she is at her best at two.

Summer worker bees have only a life expectancy of six weeks, whilst winter workers can last six months.

Drone lifespan varies and can be around four months.

Diva

The character of the queen determines that of the rest of the hive.

In order to personalise and develop the best traits in their stock, beekeepers will usually begin queen rearing for themselves.

Feet

One of the special pheromones a queen carries is in her feet.

The oily residue can help worker bees to track her down anywhere.

Golf

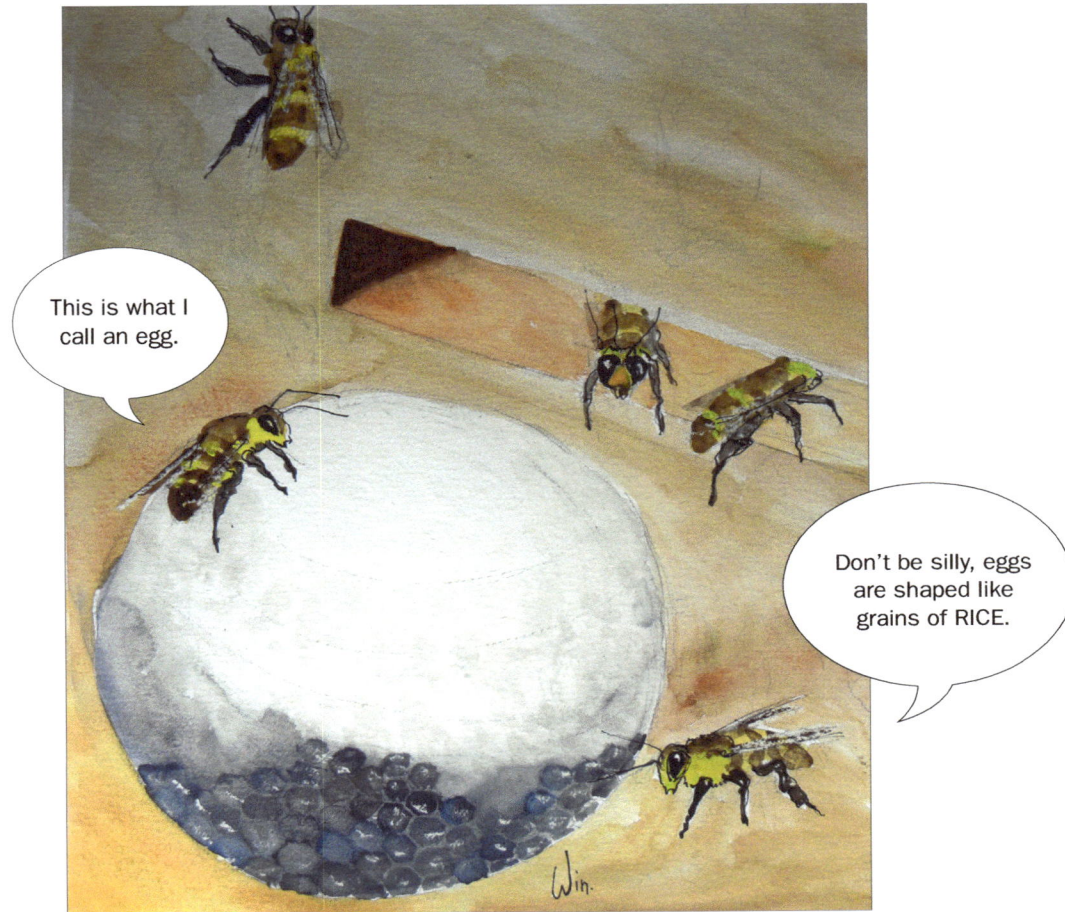

Honeybee eggs are usually described as like "grains of rice".

They are important to see and this is sometimes difficult.

A snipped off tiny piece of white cotton may be a better description.

Hide

Finding the queen appears to be "step one" in a variety of beekeeping manoeuvres.

It is made more difficult that the queen pheromone is such an attractant to workers that they often cover her.

Cage

It is vital to have a healthy, strong queen in each hive. Occasionally, it becomes necessary to re-queen a colony; perhaps it becomes queenless or aggressive.

Sometimes a queen is necessary to reduce the effect of disease. The new queen is placed in an introduction cage, usually plugged with fondant. Once the workers have freed her, the idea is that they will be used to her scent and will accept her.

Mark

For some manipulations the queen has to be found quickly.

If she is marked with coloured ink in Spring it really helps.

Clumsy attempts by new beekeepers often end badly.

Piping

One of the most remarkable sounds in nature is that of a queen honeybee Piping.

Although not with bagpipes, queens do pipe, usually in their cells when they "Quack".

Once emerged, they "Toot". This is very loud and can be heard clearly.

These noises are said to announce their presence to other bees, especially to other queens. They are issuing a challenge to a fight to the death.

Push

If the queen fails, it could be the death of the entire colony.

Bees will sense this and begin to raise a new queen to replace the old.

If the colony is large, they may decide to swarm.

To maintain young vigorous queens, it is wise to replace them after their second year.

Queen Cell

Ideally, there can only be one, strong, fertile queen in a hive.

Usually several "peanut-shaped" queen cells are built. The first emerged queen
will find the rest, tear open the side of the cell and dispatch her rivals.

Survival of the fittest!

Virgin

Occasionally there can be two queens or more in a hive but this is only for a short time.

Once a queen emerges, she will destroy any other queen
cells she can find or fight a rival to the death.

Just before her mating flight, it is known for a young queen to live
peacefully for a time with her mother who will eventually swarm.

Wing

Swarming is the natural way of splitting the colony.

It is not something a beekeeper wants to happen, as often half the colony, plus a good queen, is lost.

There are several ways to prevent this happening. Some beekeepers will clip the
queens wing, preventing her from flying and thus leading the swarm.

Chapter 3

Drones

> **"** *It doesn't matter who my father was; it matters who I remember he was.* **"**

Anne Sexton

Congregation

In the mating season, drones will fly into congregations in the sky. Virgin queens will find these groups and mating will take place on the wing.

Drifting

Drones do not tend to stay in the hive from where they emerged. They tend to drift.

Sometimes a single drone will find food and shelter in several other hives. Sadly, this is one of the ways pests and disease is spread between colonies.

Workers usually freely accept males until Autumn when drones are no longer needed for mating. They are then evicted from all hives.

Father

The drone honeybee has no father. He develops from an unfertilised
egg and only has the mothers set of genes (he is haploid).

In some circumstances it is possible for workers to begin laying. As they cannot
mate, they only produce drones as does a queen who has not mated properly.

Neither of these scenarios is welcome to the beekeeper.

Job

Unlike the many roles of the workers the drone has one main role and that is to mate with a queen. His role may be small but it is nevertheless vital to the survival of the hive.

Drone

A queen bee will usually leave the hive once or twice in her lifetime.

She will mate on the wing with up to 20 male bees (drones).

She will find them in a congregation in the sky.

Evict

As the weather grows colder each year, the drones die off or are driven from the hives.

Winter bees are all female workers who surround, feed and keep
the temperature up around the queen in a huddle.

Young new drones will appear again in the Spring, ready to mate with new virgin queens.

Fly

Drones have the strongest wing muscles of all honeybees. They will fly several miles to find a drone congregation area and then will still be able to summon up the burst of speed necessary to catch and mate with a queen.

Gamete

The term "Flying gamete" is often used to describe the drone. It is actually quite accurate. His main role is to make sure that his genetic material is passed on to the next generation.

The fact that the queen will mate with up to 20 ensures a good variety of characteristics which will serve to strengthen the hive.

Mated

A queen could mate with up to 20 drones on her mating flight. Each one will die from the experience.

Although the father dies, the queen may go on to produce many of his offspring

Pheromone

Queen honeybees produce several types of pheromones from different parts of their bodies. They enable other bees to find and recognise her also diagnosing her physical state. They efficiently enable drones to track her over many miles in order to mate.

Sting

Only the queen and the worker bees have stings. The worker stings are barbed and will remain in the skin of a human, killing the bee.

The queens sting is not barbed but long and thin, designed for injecting into queen cells, thus killing her rivals.

The drone, with his rounded tail does not have a sting and relies on the protection of the workers.

Tagged

With new technology wildlife can be tracked in the name of scientific studies.

Honeybee research is using tracking devices to monitor various behaviours e.g. drone drifting into several hives and drone congregation areas during mating.

Chapter 4

Behaviour

> **"** *A man's ethical behaviour should be based effectually on sympathy, education and social ties and needs.* **"**

Albert Einstein

Ball

One of the honeybees most effective defence strategies is to "Ball": a group effort of stinging and surrounding to smother the victim.

Canola

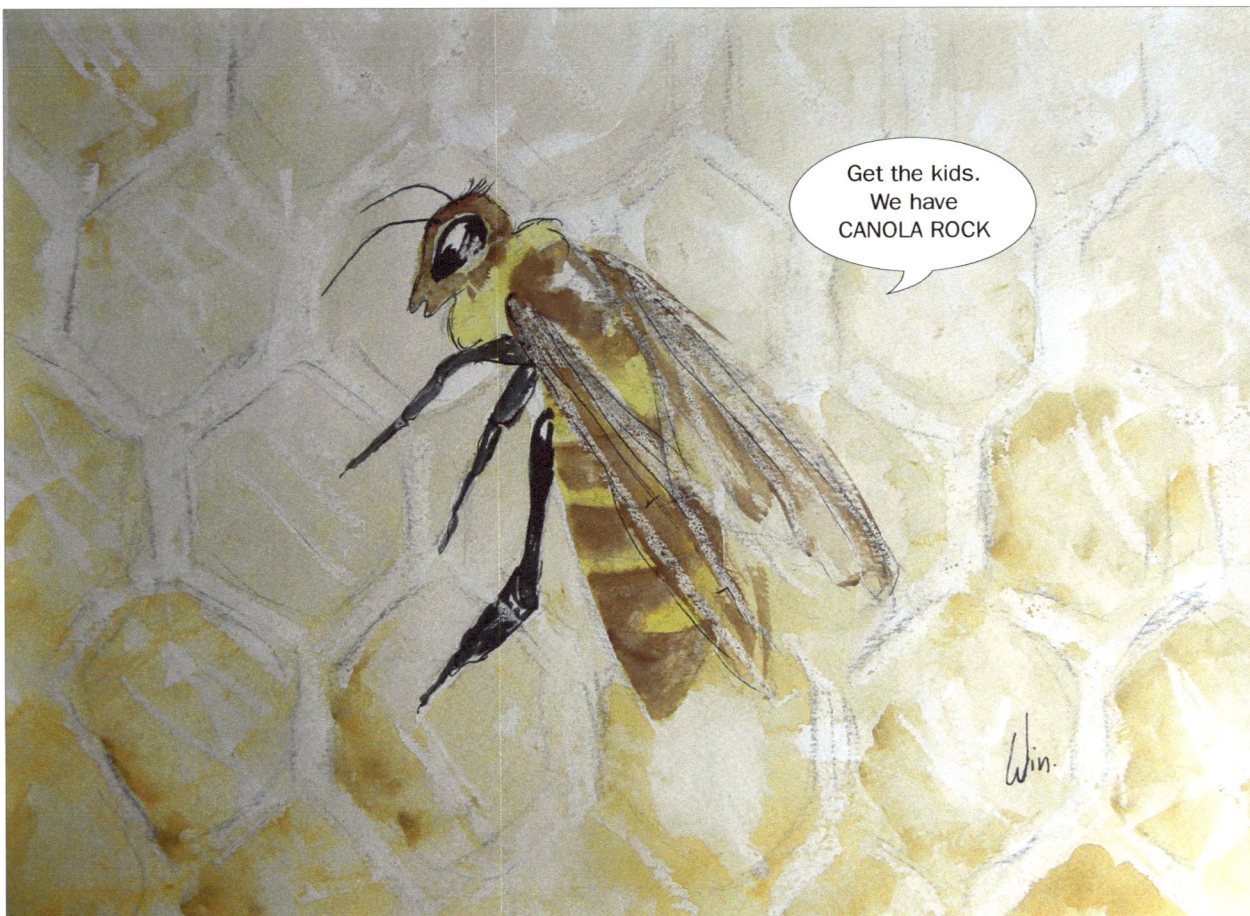

Canola, also known as oilseed rape, can be a wonderful crop for beekeepers.

It grows profusely and can give a bumper crop of honey in the Spring.

On the downside, it will granulate at a rapid rate, making it very difficult to extract.

A prompt removal by a vigilant beekeeper is what is required.

Dance

The many types of "dance" performed by a honeybee are still being researched.
They have a variety of informative purposes e.g. telling others to pay attention,
food sharing, recruitment of storage or receiver bees, eliciting grooming.

Dances are also done to pinpoint the exact location of a good
food source, saving other bees time and effort.

The dancing bee usually gives out samples of the food which is being recommended.

Dark

It is difficult to imagine how a honeybee can perform such complex manoeuvres in the darkness of the hive.

The honey bee spends most of its life performing highly intricate, complicated tasks. Most of this is done in the pitch blackness of a crowded hive.

Death

Foraging, the most demanding of activities, is usually the final task for a honeybee.

In most cases the bee will die, worn out away from the hive.

In this way there will be no need for others to remove dead bees
which could cause infection to the rest of the hive.

Jelly

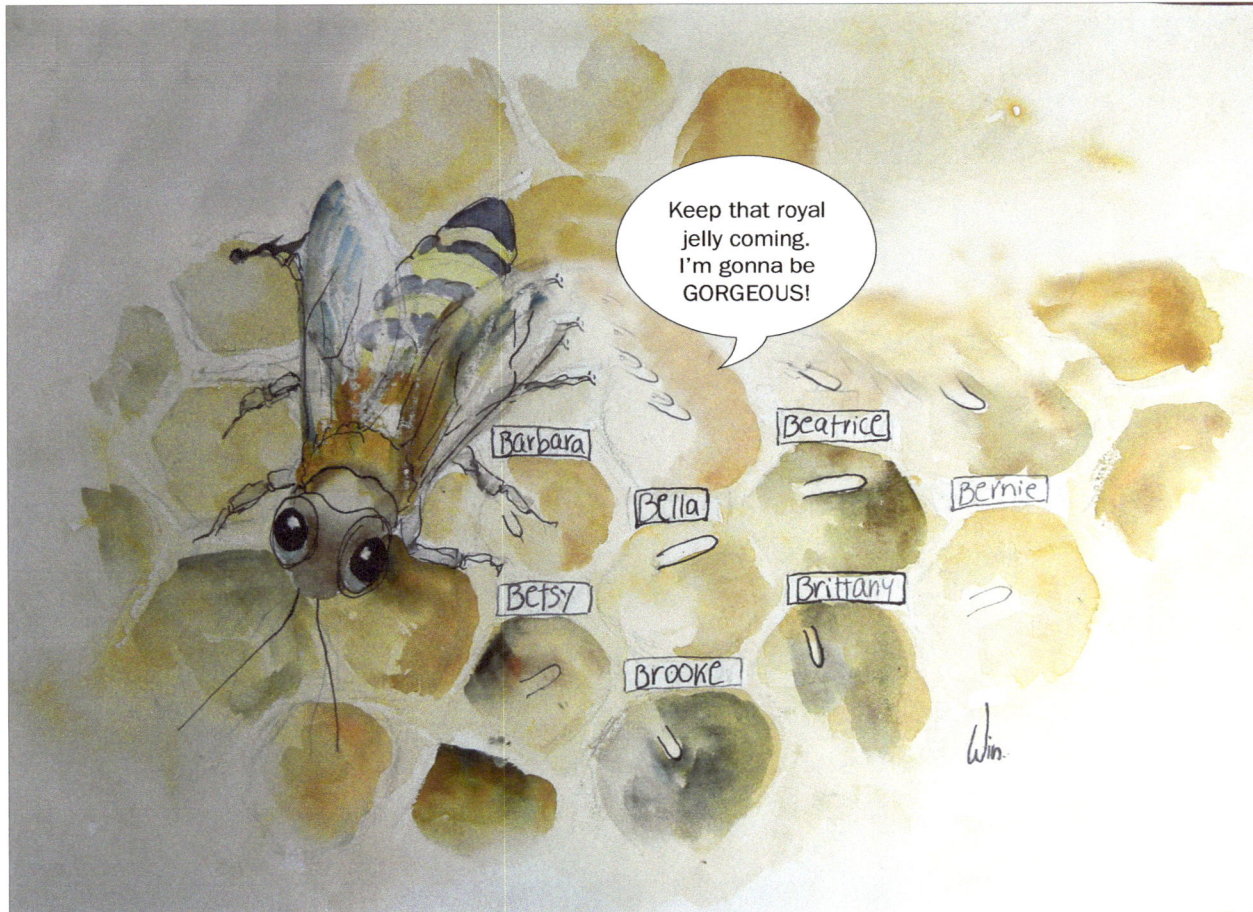

Royal jelly is the substance made by worker bees and fed to all larvae.

The diet of workers is changed to bee bread, a mixture of pollen and nectar
but the larvae destined to be queens are kept on the jelly.

This has also become very popular in the beauty industry, hence the reference to becoming gorgeous.

Nazonov

The nazonov gland is situated above the last tergite (segment) of a honeybee, it can be seen as a white flash when the tail is lifted. The pheromone released from it and fanned will attract other bees. This is useful when the colony is separated as in a swarm or hive manipulation.

It is said to have an odour similar to bananas. Bees have a strong sense of smell and it may be wise not to eat bananas or spray fragrance before inspecting a hive.

Peroxide

Bees have to make several chemical changes to the nectar to covert it into honey.

They evaporate the water content to 17-18%. They add sucrose (from their hypopharangeal glands) and also from here they add glucose oxidase. This breaks down the nectar into gluconic acid and bacteria destroying hydrogen peroxide which also happens to be a constituent of human hair bleach.

Propolis

Bee products include not only honey and wax but the wonderful substance propolis. It is used by the bees for its anti bacterial properties. Honeycomb cells are lined with it, tiny gaps are filled with it and large objects which cannot be removed are covered with it.

The name comes from Greek. The literal translation is "before the city". Bees can use it to help make the entrance smaller. Its medicinal properties in humans are extensive but beekeepers would often rather do without it, as the sticky nature makes their job harder.

Swarm

In order that the queen can lead a swarm, the workers will stop feeding her. She will then lose weight and be able to fly once more. The only other occasions will have been her mating flights.

Veil

Defensive bees will find any little chink in a beekeepers armour. It pays to spend an extra few minutes checking that everything is fully closed.

Once one bee has stung, a pheromone is released which encourages other bees to sting in the same spot. A sting should be removed at once and the area puffed with the smoker to mask the odour.

Wet

Temperature is not a problem for bees. They can regulate heat or cold by fanning or shivering.

Wetness in the hive can be an immense problem. Beekeepers should site
hives in a sheltered spot but not under trees to avoid drips.

Damaged equipment can be a cause of damp hives and colony loss.
This has to be regularly checked, mended or replaced.

Chapter 5

Pests and Diseases

> **"** *I believe it is better to be prepared for an illness than to wait for a cure.* **"**
>
> Roger Moore

Healthy Larvae

Many of the nastiest honeybee diseases can be found in the brood of a hive. A beekeeper soon learns to recognise any symptoms and knows how to deal with them at the earliest juncture.

Hygienic

In an attempt to combat pests and diseases, particularly the varroa mite, special queens are being bred. They are selected for hygienic grooming behaviour, which in turn is passed on to the workers. Evidence is showing that this is a successful and natural method.

Mice

Unwelcome objects in the hive are simply removed by the worker bees.
If an object is too large to carry it is covered in propolis.

Mice can enter a hive to make a home for the winter. They can cause a great deal of damage to the hive and its combs. They will scratch holes and pollute the colony with urine. A strong colony will kill them and propolise the body. A beekeeper can help by fitting a mouseguard to the hive in Autumn.

Missing

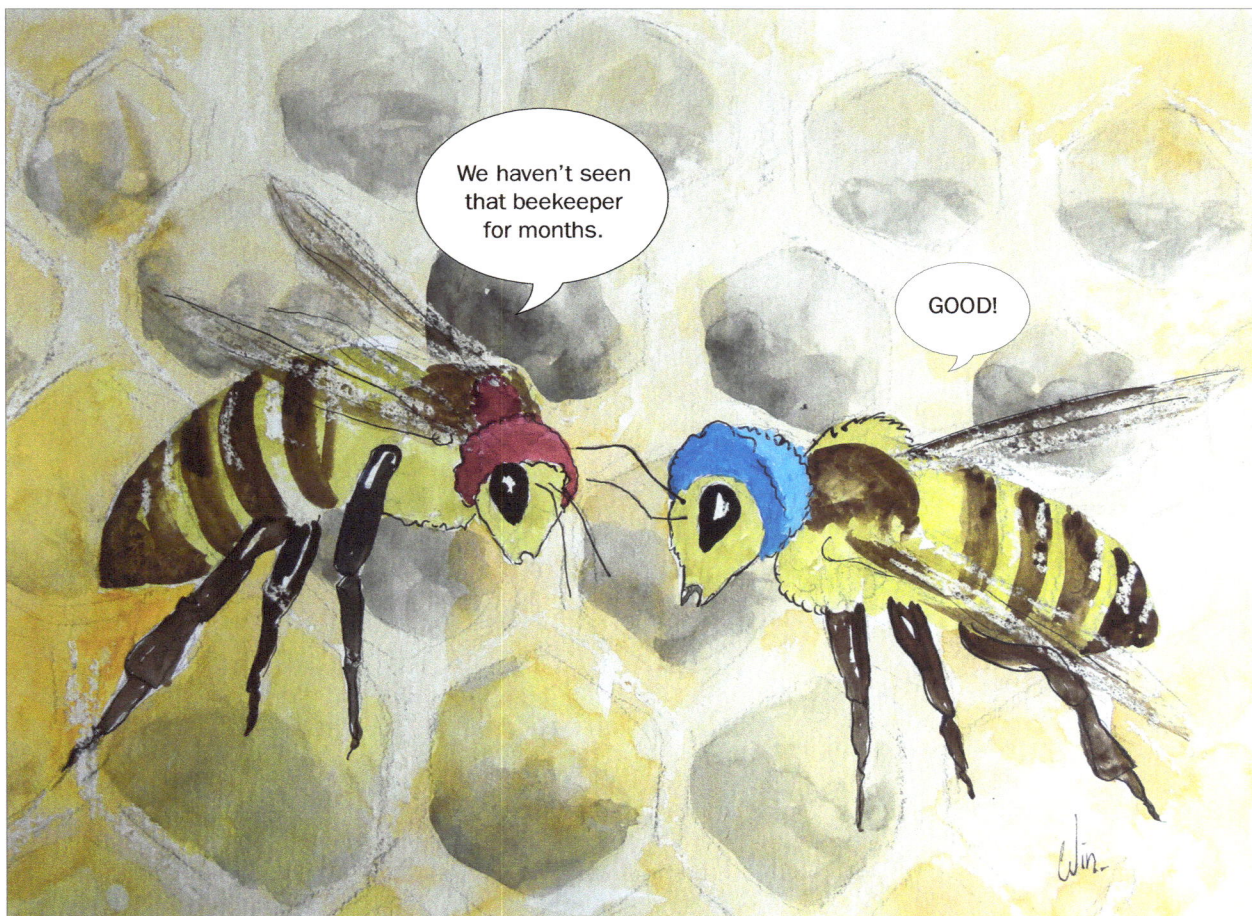

In winter, it would not be wise to continue to go into hives. Bees struggle to keep temperatures up. Apiary visits can be kept to a minimum. A hive can be hefted (gently lifted on one side)

If it is light a pollen patty or fondant can be quickly placed on the top bars.
Even in summer beekeepers can over-visit. Once a week in swarm season is
enough and then with a definite plan as to what needs to be done

Mites

Honeybees emerge at different times.

Queens take 15-16 days, workers 21 days, whilst the larger drones take a full 24 days.

This is why varroa mites usually prefer a drone host, as they can feed on a larvae for longer.

Neonicotinoids

The evidence is mounting to prove that pesticides containing neonicotinoids are harmful to bees.

Damage ranges in severity from total colony collapse to bees unable to communicate or navigate back to the hive. Surely it is only a matter of time before products like this have a worldwide ban.

Nosema

Nosema is a bacterial disease which attacks the gut of the bee.

It can cause dysentery and may also destroy a colony.

Feeding bees the wrong sort of food – e.g. brown sugar – can also cause dysentery.

SHB

At the time of writing this particularly nasty pest is not thought to be in the British Isles. It is however a major threat to colonies in America, Canada, Mexico, Australia and point of source, Africa.

Infestations of SHB larvae will destroy comb and pollute honey. It reduces the hive to a mass of smelly slime causing colonies to abscond or die.

There are several treatments being developed, chemical strips, nematodes and traps. Colony numbers should be kept high and free comb reduced to a minimum.

Story

"Oh Grandma", exclaimed Little Red Riding Hood. "What big yellow legs you have".

One of the earliest tasks of a honeybee worker is that of a nurse. She will prepare the cells for laying. Once laid she will feed the eggs and larvae, finally capping each cell to await the emerging bee.

They are not read fairy stories about scary monsters but the one referred to in this cartoon is the yellow-legged Asian hornet.

Varroa

Beekeeping must have been idyllic pre 1992 when varroa was found in the UK.

Total "leave alone" beekeeping has now to be reconsidered. With a little experience, the most natural interventions can be used to control this nasty little parasite which can be a vector for many unpleasant diseases. Many of them disabling and some fatal.

Wasps

In Autumn. carnivorous wasps are attracted to the bees and their honey.

When the hive should be at its strongest, preparing for Winter,
wasp attacks can cause the destruction of a colony.

A wise beekeeper could help by reducing hive entrance size with
foam or tape. Wasp traps in the apiary are also useful.

Wax Moth

The wax moth is a pest whose larvae can decimate the comb of a hive in a very short time.

To reduce risk of infection make sure all hive parts are hygienically clean by scrubbing with washing soda and blasting with a heat gun. Steaming is also an option.

Empty hives should have entrances tapped. Some people pop a sprig of wormwood in as a deterrent.

Chapter 6

Beekeeping

> **"** *There are certain pursuits which, if not wholly poetic and true, do at least suggest a nobler and finer relation to nature that we know. The keeping of bees for instance.* **"**

Henry David Thoreau

Adam

This cartoon is a reference to the famous Brother Adam who was beekeeper at Buckfast Abbey for many years.

He was one of the pioneers in queen rearing and developed his own strain the "Buckfast bee" in which he attempted to incorporate the best features of bees which he found on his extensive travels.

Ash

The smoker is a tool which is said to calm the bees. Experienced beekeepers use very little. One of the first things a new beekeeper learns is how to light one and keep it lit. The fuel is very important. A good mix of fast and slow burning natural materials works well.

Beekeeper

Sadly, the worst enemy of a honeybee is a bad beekeeper.

Careless handling, maintenance and replacement of hive parts kills bees.
Poor hygiene causes disease to be spread between colonies. Ignorance
in the diagnosis and treatment of disease kills entire colonies.

Honey is taken and bees are left to starve unless some form of replacement is given.

Bumble

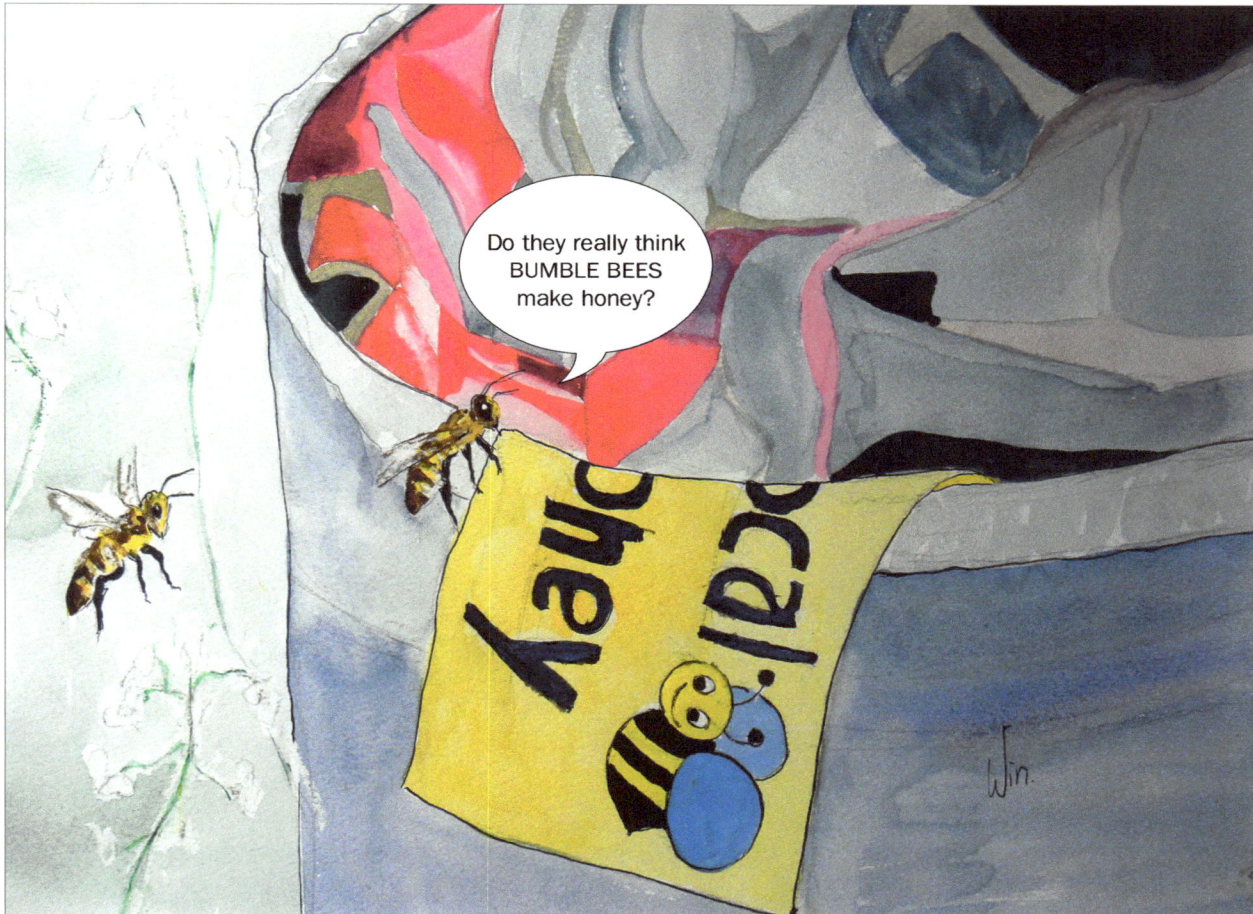

Many people do not know what a honey bee looks like. They assume it is the rounded bee seen on too many posters and honey labels. This is more like a bumble bee.

Beekeepers are regularly called out to swarms of bumble bees as a result of this. Honeybees are in fact small, elegant and very pretty.

Cake

In the depth of Winter, usually around Christmas time, many beekeepers give fondant to the bees. The block is placed directly above the cluster of bees. It is easily assimilated and reached by bees which otherwise may starve.

Contact

The contact feeder is probably the simplest feeder on the market. It is basically a container with holes or fine mesh on the lid. Once filled with sugar syrup, the lid is fitted securely and the device is inverted over a hole in the crown board. This should be held by a vacuum but it is wise to invert over a spare bucket before lowering on to the hive.

Mite Strips

Varroa treatments can sometimes be really smelly. Some people prefer not to treat.

When necessary, a treatment with a preparatory, natural product can protect bees from a variety of horrendous diseases or sometimes total colony collapse.

Pollen

At the end of Winter or the very start of Spring the weather may still be too poor for bees to forage.

Pollen is the main source of bee protein and the addition of a pollen pattie in these conditions may save the life of a hive and perhaps encourage early egg laying.

Rapid

In the early (or later) months of the year, rapid feeders are a great way to get food into a weak colony.

All parts have to be secured as quickly as possible as inevitably
several bees will end up floating in the syrup.

Rustlers

Hive rustling, a despicable crime, is on the increase. It is wise to take some sort of precaution when setting up an apiary. Permanently marking hives is a good idea, as is CCTV.

Syrup

Feeding bees when necessary could be a matter of life an death.

Sugar and water syrup is given in a variety of feeders. Usually the colder the weather, the thicker the solution. In winter 2kg of sugar to 1 pint of water is good. In Spring this becomes 2kg to 2 pints, hence the "watery" comment.

Some beekeepers like to add any treatments to the syrup. Some of these are thymol-based.

Topbar

The topbar hive is one of the oldest hive designs.

It is on a single level. No foundation frames are used and the bees are allowed to draw their own wild comb.

www.ingramcontent.com/pod-product-compliance
Lightning Source LLC
Chambersburg PA
CBHW041300210326

41599CB00006B/252